Quantum Computing
Schrödinger's Code

How the Strangest Science in the World Outperforms the Machines of Today

Joe E. Grayson

Copyright © 2024 Joe E. Grayson, All rights reserved.

No part of this publication may be reproduced, distributed, or transmitted in any form or by any means, including photocopying, recording, or other electronic or mechanical methods, without the prior written permission of the publisher, except in the case of brief quotations embodied in critical reviews and certain other noncommercial uses permitted by copyright law.

Table of Contents

Table of Contents

Introduction

Chapter 1: The Genesis of Quantum Computing

Chapter 2: Foundations of Quantum Computing

Chapter 3: The Power of Quantum Computing

Chapter 4: Real-Life Demonstrations

Chapter 5: Practical Applications

Chapter 6: Challenges and Limitations

Chapter 7: The Future of Quantum Computing

Chapter 8: Quantum Mysteries: Probing the Universe

Conclusion

Introduction

In the vast realm of technological innovation, few advancements have captured the imagination of scientists, futurists, and the general public quite like quantum computing. Its potential to redefine what is computationally possible has placed it at the forefront of cutting-edge science and technology. Unlike the step-by-step evolution of traditional computing, quantum computing represents a revolutionary leap—one that doesn't just push the boundaries but redefines them entirely. It is a bold promise of transformation, not merely an enhancement of existing capabilities but the creation of something fundamentally new.

Imagine a computer that operates on the principles of the very fabric of reality—one that leverages the enigmatic behaviors of subatomic particles to solve problems that would take conventional computers millions of years. Quantum computing stands apart from any other breakthrough because it doesn't aim to refine the tools we already have; instead, it introduces a paradigm that challenges our fundamental understanding of computation itself.

A useful way to grasp the strange but fascinating principles of quantum computing is to consider the famous thought experiment of Schrödinger's cat. Picture a cat in a sealed box with a vial of poison, a radioactive particle, and a mechanism that will release the poison if the particle decays. According to quantum mechanics, until the box is opened and observed, the cat exists in a state of superposition—it is both alive and dead at the same time. This paradoxical state, where two seemingly contradictory outcomes coexist, is at the heart of quantum computing. Instead of processing information as binary bits—ones and zeros—quantum computers operate using quantum bits, or qubits, that can represent a one, a zero, or both simultaneously.

This strange and counterintuitive principle is not a flaw but a feature, and it unlocks possibilities that go far beyond the scope of classical machines. Through superposition, entanglement, and interference, quantum computers have the potential to solve problems that are practically impossible for today's most advanced supercomputers. From unraveling the complexities of molecular structures for drug development to cracking encryption codes and optimizing massive systems, the promise of quantum computing stretches across industries and disciplines.

As we embark on this exploration, the journey will unveil the foundational concepts that make quantum computing possible, from its use of qubits to the unique properties that give it such extraordinary power. We will also delve into the practical applications already taking shape, the challenges that stand in the way of widespread adoption, and the profound implications this technology holds for our future. With quantum computing, we are not just building faster machines—we are venturing into the fabric of reality itself, where the possibilities are as vast and uncertain as the universe it seeks to decode.

Chapter 1: The Genesis of Quantum Computing

The story of modern technology is deeply intertwined with the rise of classical computing. These machines, powered by binary logic, transformed how humanity processes information, solves problems, and connects across the globe. At the heart of classical computing lies a simple yet powerful foundation: the bit. These tiny switches, represented as either a one or a zero, serve as the building blocks of computation. By combining billions of these bits into intricate circuits and leveraging algorithms, classical computers have revolutionized industries, enabling everything from real-time global communication to intricate simulations of the cosmos.

Yet, despite their transformative impact, classical computers face a fundamental limitation. They excel at tasks where the steps are sequential and predictable but falter when confronted with problems that grow exponentially in complexity. For instance, while a classical computer can efficiently calculate a multiplication of two numbers, breaking the result into its prime factors—a cornerstone of modern encryption—becomes an insurmountable task as the numbers grow larger. These kinds of challenges expose the boundaries of classical computation, where time and resources expand exponentially as problems increase in size.

The constraints of classical machines have led scientists to seek alternatives, which brings us to the curious world of quantum mechanics—a domain that defies intuition and operates on principles fundamentally different from those of classical physics. Quantum mechanics emerged in the early 20th century as scientists probed the behaviors of the smallest building blocks of reality: atoms and subatomic particles. Unlike the deterministic rules of classical mechanics, the quantum realm is governed by probabilities and uncertainties, creating a landscape of possibilities that seemed paradoxical.

To illustrate this strange world, the physicist Erwin Schrödinger introduced his now-famous thought experiment involving a cat. In this scenario, a cat placed in a sealed box with a quantum mechanism could be simultaneously alive and dead, existing in a state of superposition until observed. While this paradox was initially a way to highlight the oddities of quantum mechanics, it also became a powerful metaphor for understanding the principles that underpin quantum computing.

At the core of quantum mechanics are phenomena like superposition and entanglement. Superposition allows particles—or in the case of quantum computing, qubits—to exist in multiple states at once. Unlike a classical bit that is confined to a single value of either zero or one, a qubit can represent a spectrum of probabilities, opening up exponentially more possibilities. Entanglement, another hallmark of the quantum world, links particles in such a way that the state of one instantaneously affects the state of another, even across great distances. This interconnectedness creates pathways for information that defy conventional limitations.

By harnessing these quantum properties, researchers have opened a door to an entirely new approach to computation. Quantum computing doesn't just aim to improve upon the tools of classical systems—it reimagines the very process of computation itself, offering the potential to tackle challenges previously thought unsolvable. It is in this convergence of the strange and the possible that quantum computing finds its origin and promise.

Chapter 2: Foundations of Quantum Computing

3.1. In the world of classical computing, everything begins with bits. These fundamental units of information operate as binary switches, taking on values of either 0 or 1. This simplicity is the cornerstone of classical computers, allowing them to process instructions sequentially, store data, and execute algorithms. By combining billions of these bits into sophisticated architectures, modern computers can achieve astonishing feats, from real-time communication to processing massive datasets. However, their binary nature also imposes limits. A bit can only ever exist as one value at a time—either 0 or 1—leaving classical computers confined to a linear method of computation.

Enter the qubit, the quantum counterpart to the classical bit. Unlike its binary cousin, a qubit exists not as a discrete state but as a blend of possibilities. This property, known as superposition, allows a qubit to represent 0, 1, or any combination of the two simultaneously. Imagine spinning a coin in midair: until it lands, it exists in a state that is neither heads nor tails but a mixture of both. Similarly, a qubit holds a range of probabilities, vastly expanding the computational possibilities. When qubits are measured, their probabilistic nature comes into play, collapsing into a definite state of either 0 or 1 based on their likelihood.

Manipulating qubits requires a different set of tools than those used for classical bits. Quantum gates serve as these tools, functioning like classical logic gates but with unique capabilities. For example, the Hadamard gate is pivotal in quantum computing, as it creates superposition by transforming a qubit from a defined state into a balanced blend of possibilities. By stringing together multiple gates, researchers build quantum circuits that manipulate qubits in precise and controlled ways, laying the groundwork for powerful computations.

The magic of quantum computing doesn't stop with superposition. Quantum interference, another remarkable property, allows operations to amplify correct solutions while canceling out incorrect ones. This process underpins the ability of quantum computers to solve certain problems exponentially faster than classical systems. Through parallelism, where a single operation can address multiple states simultaneously, quantum computers can perform many calculations at once, leveraging the probabilistic power of qubits to explore vast solution spaces.

Perhaps the most intriguing aspect of quantum computing is entanglement, a phenomenon so strange that Einstein famously referred to it as "spooky action at a distance." When two qubits become entangled, their states are inextricably linked, no matter how far apart they are. A change in one qubit instantaneously affects the other, enabling a level of correlation and communication that defies classical understanding. For instance, two entangled qubits can exist in a state where they are guaranteed to show identical results when measured, whether both are 0 or both are 1, while never producing mismatched outcomes. This interconnectedness lays the foundation for revolutionary capabilities like quantum teleportation and advanced cryptographic systems.

These unique properties—superposition, interference, and entanglement—collectively redefine what computers can achieve. By stepping beyond the binary constraints of classical bits and leveraging the enigmatic behaviors of qubits, quantum computing opens doors to a world of possibilities that traditional systems could never explore. It is a new frontier, as much about understanding the universe as it is about harnessing its principles for transformative technology.

Chapter 3: The Power of Quantum Computing

Quantum computers hold the potential to revolutionize the world of computation, offering speeds and capabilities that far exceed those of classical machines. While traditional computers rely on bits to process information, quantum computers use qubits, which can exist in a superposition of states, allowing them to perform complex calculations exponentially faster. This fundamental difference opens up entirely new avenues for solving problems that would be nearly impossible for classical computers.

One of the most promising areas where quantum computers can outshine classical machines is in solving encryption algorithms. Modern encryption methods, like RSA, are built on the difficulty of factoring large numbers. Classical computers take an enormous amount of time to break these codes, often relying on brute-force methods that take millions of years to complete. In contrast, quantum computers can harness algorithms like **Shor's algorithm**, which efficiently factor large numbers by exploiting quantum parallelism and interference. This means that a quantum computer could theoretically break encryption in a fraction of the time it would take a classical computer, disrupting the current security landscape but also paving the way for unbreakable quantum encryption systems.

Beyond encryption, quantum computers are poised to make significant strides in fields like **drug discovery**. Classical computers simulate molecular interactions by solving complex equations, but these simulations are limited by the sheer number of variables involved. Quantum computers, however, operate on the same quantum principles that govern molecules, allowing them to simulate the behavior of atoms and particles with unprecedented accuracy. For instance, simulating the interactions of proteins and molecules can reveal insights into how diseases like cancer or Alzheimer's develop, offering the potential to design drugs that can target these conditions more effectively. The ability to perform these simulations more efficiently could drastically shorten the time needed for drug discovery and lead to breakthroughs in personalized medicine.

Finally, quantum computing shows immense promise in **optimizing large, complex systems** such as financial portfolios and supply chains.

Traditional optimization methods struggle to handle the vast number of variables and scenarios involved in such systems. Quantum computers, however, can perform optimizations using quantum algorithms like **the Quantum Approximate Optimization Algorithm (QAOA)**. These algorithms enable quantum machines to explore multiple possible solutions simultaneously, finding the best possible outcome in a fraction of the time. In finance, this could mean real-time risk analysis and portfolio optimization on a scale never before possible, while in logistics, it could lead to more efficient supply chains, reducing costs and waste on a global scale.

In all these fields, the unique properties of quantum computing—such as superposition, entanglement, and quantum interference—empower it to outperform classical computers in solving problems that would take traditional machines an impractically long time. The potential of quantum computing to revolutionize encryption, drug discovery, and optimization is just the beginning. As quantum hardware continues to evolve, we are only beginning to scratch the surface of what's possible with this new form of computation.

Chapter 4: Real-Life Demonstrations

One of the simplest yet most striking demonstrations of a quantum computer's capabilities is the coin game experiment. Imagine a coin game where a human player competes against a computer. The rules are straightforward: the coin starts on heads, and each player, starting with the computer, can either flip the coin or leave it unchanged. However, neither player gets to see the coin's state after each turn. The game ends after three rounds, and if the coin shows heads, the computer wins; if it's tails, the human wins.

In a classical scenario, this game would be fair, offering both players a 50% chance of winning. The coin flips and decisions would result in randomness, with neither player holding an inherent advantage. However, when the same game is played against a quantum computer, the results defy expectation. In an experiment conducted with IBM's quantum computer, the quantum machine won nearly every game. How did it achieve this seemingly magical outcome?

The secret lies in the quantum computer's ability to maintain the coin in a state of superposition. Rather than committing to a fixed state of heads or tails, the quantum computer effectively kept the coin in a probabilistic blend of both outcomes throughout the game. This meant that no matter what action the human player took—flipping the coin or leaving it unchanged—the quantum computer could preserve the superposition. In its final move, the quantum computer "unmixed" the state, ensuring the coin landed on heads. This demonstration highlights the profound implications of quantum mechanics, where superposition and interference can be leveraged to guarantee outcomes that would otherwise be governed by chance.

The principles demonstrated in the coin game extend far beyond playful experiments. In research labs around the world, quantum computers are being used to simulate phenomena that challenge classical understanding. One of the most exciting areas of experimentation is **quantum teleportation**. By harnessing the phenomenon of entanglement, scientists have successfully transmitted the state of one quantum particle to another, even across significant distances. This process doesn't involve physically moving the particle but instead allows its quantum state to be recreated in another location. While teleportation of physical objects remains in the realm of science fiction, this technique holds immense promise for secure and instantaneous data transmission.

Building on this, quantum researchers are also exploring ways to create secure communication networks. Quantum mechanics introduces an inherent layer of security: any attempt to eavesdrop on a quantum communication channel would disturb the system, making the intrusion immediately detectable. Experiments in quantum encryption and data transfer have already demonstrated the feasibility of creating unbreakable communication systems, laying the groundwork for what could become a global quantum internet.

Both the coin game and these groundbreaking lab experiments underscore the unique power of quantum computing. While the coin game offers a glimpse into the counterintuitive principles of quantum mechanics, real-world applications like teleportation and secure data transmission reveal the profound impact quantum technologies could have on communication and information security. These achievements are not just theoretical exercises—they represent tangible steps toward a quantum-enabled future.

Chapter 5: Practical Applications

Quantum computing offers revolutionary advances in areas where classical systems face limitations, and one of its most transformative promises lies in the realm of encryption. Traditional encryption methods rely on the computational difficulty of specific mathematical problems, such as factoring large numbers or solving discrete logarithms. While these techniques have secured our digital world for decades, they are vulnerable to quantum algorithms like Shor's algorithm, which can unravel such protections with unprecedented speed. In response, quantum encryption offers a solution rooted in the laws of quantum mechanics themselves, ensuring a level of security that cannot be breached without fundamentally breaking the principles of physics.

At the heart of quantum encryption lies the concept of **quantum key distribution (QKD)**, which utilizes the inherent uncertainty of quantum states. When a quantum key is transmitted, any attempt to intercept or measure the key alters its state, alerting the sender and receiver to the intrusion. This feature makes quantum encryption theoretically unbreakable, as eavesdroppers cannot access the key without leaving traces of their presence. Banks, government agencies, and corporations are already experimenting with QKD to secure communications, and as the technology matures, it could redefine cybersecurity for the modern era. Beyond banking and communications, quantum encryption could safeguard sensitive medical data, financial transactions, and even military communications, creating an impenetrable barrier against cyber threats.

Quantum computing's potential extends beyond security, finding promising applications in healthcare and medicine. Classical computers struggle to simulate the complex quantum interactions within molecules, limiting our ability to design drugs that precisely target diseases. Quantum computers, operating on the same principles as these molecular interactions, are uniquely equipped to tackle these challenges. They can simulate molecular structures, predict how different compounds will interact, and identify the most effective candidates for drug development—all in a fraction of the time required by classical methods.

For example, researchers are exploring how quantum simulations could accelerate the search for treatments for diseases like Alzheimer's, which affects millions of people worldwide. By modeling the intricate folding patterns of proteins associated with the disease, quantum computers could reveal vulnerabilities that are invisible to classical approaches. This capability could lead to breakthroughs in treating not only Alzheimer's but also cancers, genetic disorders, and infectious diseases, revolutionizing the field of medicine and improving countless lives.

Another groundbreaking frontier is the development of the **quantum internet**, a network that leverages quantum entanglement to enable instant and secure data transfer. Unlike classical communication networks, which rely on physical transmission of data, the quantum internet uses entangled particles to share information across vast distances. When one particle's state is changed, its entangled partner reflects the change instantaneously, creating a channel for data transfer that is not only faster but also immune to traditional forms of hacking.

Research into building a quantum internet is already underway, with experimental quantum networks demonstrating secure communication between nodes. These systems rely on entanglement-based protocols to ensure that any interception attempt disrupts the entangled states, making breaches detectable. As this technology evolves, it holds the potential to transform global communication, enabling secure voting systems, unhackable financial transactions, and even real-time collaboration across continents.

The possibilities offered by quantum encryption, healthcare simulations, and the quantum internet are more than just theoretical. They represent a profound shift in how we protect, heal, and connect in an increasingly complex world. By harnessing the strange and powerful principles of quantum mechanics, these technologies promise a future where information is secure, diseases are more treatable, and communication transcends traditional limits.

Chapter 6: Challenges and Limitations

While quantum computing holds immense promise, the path to its widespread adoption is riddled with significant challenges, particularly in the realm of hardware. At the heart of these challenges lie the operational errors and noise inherent in quantum systems. Qubits, the fundamental units of quantum computation, are incredibly fragile. They are highly sensitive to their environment, with even minute disturbances like temperature fluctuations, electromagnetic interference, or cosmic rays disrupting their delicate quantum states. This instability, known as "decoherence," causes qubits to lose their information rapidly, limiting the reliability and precision of quantum computations.

Adding to this complexity is the issue of noise. Quantum systems are not yet precise enough to perform long computations without introducing errors. As operations accumulate, so do these errors, compounding the difficulty of achieving accurate results. Error correction, a critical component in classical computing, is far more complicated in the quantum realm due to the probabilistic nature of qubits. While researchers have developed techniques for quantum error correction, these methods require a significant overhead of additional qubits, further straining already limited resources.

Scalability is another major hurdle. Building a quantum computer that can outperform classical machines, known as achieving "quantum advantage," demands an exponential increase in the number of high-quality, error-resistant qubits. Current quantum processors are still in their infancy, with only tens to hundreds of qubits, far below the thousands or millions needed for truly practical applications. Expanding quantum systems to this scale presents formidable engineering and technological challenges, from designing more robust qubits to creating infrastructure capable of maintaining their stability.

Cost is an equally daunting obstacle. Quantum computers require highly specialized environments, such as ultra-low temperatures near absolute zero, to maintain the quantum states of their qubits. The infrastructure to achieve and sustain these conditions is both complex and expensive, limiting access to a handful of well-funded institutions and corporations. This financial barrier slows the pace of innovation and the democratization of quantum technology, leaving it largely in the hands of a select few.

Beyond these technical and financial hurdles lies the gap between theoretical advancements and real-world applications. While quantum algorithms have demonstrated their potential on paper and in small-scale experiments, applying them to meaningful problems remains a work in progress. For instance, Shor's algorithm theoretically enables quantum computers to break encryption, but implementing it on a large enough scale to crack modern cryptographic codes is still out of reach. Similarly, while quantum simulations show promise in drug discovery and materials science, the current hardware lacks the computational power to tackle the full complexity of these tasks.

This gap highlights a critical reality: quantum computing, while transformative in its potential, is still a nascent field. Bridging the divide between theory and practical application will require not only breakthroughs in hardware but also advancements in software, error correction, and algorithm development. As researchers continue to tackle these challenges, the journey toward a functional and accessible quantum computer will likely be incremental, marked by both successes and setbacks.

Despite these limitations, the progress made so far is a testament to human ingenuity and perseverance. Each step forward brings us closer to unlocking the full potential of quantum computing, transforming it from a scientific curiosity into a powerful tool that reshapes industries and solves some of the world's most pressing problems. The challenges are immense, but so is the promise of this extraordinary technology.

Chapter 7: The Future of Quantum Computing

Quantum computing stands poised to revolutionize industries on an unprecedented scale, unlocking possibilities that were once the domain of science fiction. Its ability to solve complex problems exponentially faster than classical machines opens doors in fields ranging from cryptography and healthcare to logistics, finance, and artificial intelligence. The transformative power of this technology lies in its potential to tackle challenges that classical computers find insurmountable, providing solutions that could redefine the very fabric of modern society.

In healthcare, for instance, quantum computers could accelerate drug discovery by simulating molecular interactions with unmatched precision. Treatments for diseases like Alzheimer's, cancers, and genetic disorders might emerge far more quickly, saving lives and reducing healthcare costs. In logistics and supply chain management, quantum optimization algorithms could streamline global systems, reducing waste and improving efficiency in ways that classical systems cannot match. Financial institutions could harness quantum computing to perform risk analysis and optimize portfolios in real time, managing uncertainties with a level of detail previously unattainable.

Beyond these immediate applications, quantum computing's potential extends to the realm of artificial intelligence. Current AI systems, despite their capabilities, remain limited by the computational power of classical hardware. Quantum computers, with their ability to process vast amounts of data in parallel, could supercharge AI algorithms, enabling machines to learn, reason, and adapt more effectively. This leap could usher in breakthroughs in natural language processing, autonomous systems, and decision-making tools, fundamentally altering how we interact with technology and each other.

However, with such transformative potential comes the responsibility to develop quantum computing ethically. The very power that makes quantum computing so promising also poses significant risks. For instance, its ability to break modern encryption systems could render current cybersecurity measures obsolete, exposing sensitive data and threatening the privacy of individuals and organizations alike. To mitigate these risks, researchers and policymakers must prioritize the development of quantum-resistant cryptographic methods alongside quantum technologies.

Ethical considerations also extend to the implications of quantum computing on employment and inequality. As industries adopt quantum-powered systems, the demand for highly specialized skills will grow, potentially leaving those without access to advanced education at a disadvantage. Governments, corporations, and educational institutions must work together to ensure that the benefits of quantum technology are distributed equitably, preventing a widening of the socioeconomic divide.

Speculation about the future role of quantum computing reaches even further, touching on philosophical questions about the nature of intelligence and reality itself. As quantum computers push the boundaries of artificial intelligence, they may blur the line between human and machine cognition, raising questions about accountability, bias, and the ethical use of such advanced systems. Moreover, the ability of quantum computers to simulate physical phenomena could deepen our understanding of the universe, revealing insights into the origins of life, the nature of consciousness, and the fundamental laws of existence.

In this sense, quantum computing is more than just a technological breakthrough—it is a tool for exploration, one that has the potential to reshape not only industries but also how we perceive and interact with the world. The journey ahead requires careful stewardship, balancing ambition with responsibility, and ensuring that this powerful technology is harnessed for the greater good. As we stand on the cusp of a quantum era, the decisions we make today will shape a future where the possibilities are as boundless as the questions it seeks to answer.

Chapter 8: Quantum Mysteries: Probing the Universe

Quantum computing is not just a technological marvel; it is a window into the fundamental workings of the universe. At its core, quantum computing leverages principles that govern the behavior of the smallest particles in existence—principles that defy classical understanding and challenge our notions of reality. By building machines that operate on the rules of quantum mechanics, we gain tools not only to solve practical problems but also to explore the profound mysteries of the cosmos.

One of the most fascinating aspects of quantum computing is its ability to simulate quantum systems themselves. Classical computers, despite their power, struggle to model complex quantum interactions because these systems require exponential resources to represent accurately. A quantum computer, however, operates on the same principles as the systems it seeks to simulate, making it uniquely suited for this task. This capability opens the door to unraveling phenomena that lie at the very edges of scientific understanding, from the behavior of particles in extreme conditions to the origins of the universe.

For example, quantum computers could help scientists probe the nature of black holes, the fabric of spacetime, or the behavior of particles at the Planck scale, where the laws of classical physics break down. By modeling these systems, researchers can test theories that were once confined to the realm of mathematics, providing insights into questions about the birth of the universe, the unification of fundamental forces, and the nature of dark matter and energy.

The philosophical implications of this technology are equally profound. Quantum mechanics, upon which quantum computing is built, challenges our most basic assumptions about reality. Concepts like superposition and entanglement reveal a world where objects can exist in multiple states simultaneously and where particles separated by vast distances can instantaneously affect each other. These phenomena suggest that reality is not as fixed or local as our everyday experiences would lead us to believe.

As we harness these principles in quantum computing, we are forced to confront questions that go beyond science and into the realm of philosophy. What does it mean for something to exist in a state of probability rather than certainty? How do the rules of quantum mechanics relate to consciousness and perception? Could the universe itself be a kind of quantum computation, as some physicists speculate? These questions invite us to reconsider our understanding of reality, causality, and the role of observation in shaping the world around us.

Moreover, the ethical and societal implications of quantum technology add another layer to these philosophical considerations. As we unlock the power to simulate and manipulate the universe on an unprecedented scale, we are also assuming greater responsibility for how this power is used. Quantum computing has the potential to reshape industries, economies, and even geopolitics, raising questions about fairness, equity, and the distribution of technological benefits. It challenges us to think about the intersection of power, knowledge, and morality in a world transformed by quantum insights.

In many ways, quantum computing represents humanity's quest to push beyond the limits of our understanding, using the tools of science to explore the deepest questions about existence. It is both a reflection of our ingenuity and a reminder of the vastness of what we have yet to learn. As we continue to develop this technology, we are not just building machines—we are charting a path toward understanding the universe and our place within it, one quantum state at a time.

Conclusion

Quantum computing is a journey into the heart of the unknown, a bold exploration that merges the mysteries of quantum mechanics with humanity's relentless drive for innovation. Along the way, we've encountered concepts that defy intuition—superposition, entanglement, and interference—and glimpsed their power to redefine how we compute, communicate, and understand the universe. From cracking encryption codes to simulating molecules for life-saving drugs, and from optimizing complex systems to pioneering a quantum internet, this technology holds the promise to reshape industries and solve problems once thought unsolvable.

Yet, this journey is far from over. Quantum computing remains in its infancy, its full potential still waiting to be unlocked. Current limitations in hardware, the fragility of qubits, and the challenges of scalability remind us that the path forward will not be without obstacles. But just as we have overcome the barriers of past technological revolutions, so too will we navigate the complexities of this one. Each breakthrough, no matter how small, propels us closer to a future where quantum computing transforms not just science and technology but the very way we approach and solve the problems of our world.

What makes this journey even more profound is its philosophical depth. Quantum computing is not merely a tool; it is a lens through which we can view the universe, challenging our understanding of reality and pushing the boundaries of what we believe is possible. It bridges the practical with the metaphysical, inviting us to ponder the nature of existence, the fabric of reality, and the infinite potential of human ingenuity.

The future of quantum computing is as thrilling as it is uncertain. It is a future filled with opportunities to revolutionize industries, answer age-old scientific questions, and create technologies that will shape the next century. But it is also one that demands responsibility, foresight, and collaboration to ensure that its benefits are shared equitably and its risks are mitigated.

As we stand at the threshold of the quantum era, we are called to embrace both its challenges and its possibilities. This is a moment of profound potential, one that requires curiosity, creativity, and courage. The road ahead may be uncertain, but it is precisely this uncertainty that makes it so exciting. With each step forward, we are not only building machines of unparalleled power but also crafting a legacy of exploration and discovery for generations to come.

Quantum computing is more than a technological revolution—it is a testament to humanity's ability to dream beyond limits and turn those dreams into reality. The journey is just beginning, and the possibilities are as boundless as the questions we seek to answer. Let us move forward with hope, determination, and an unshakable belief in the power of science to illuminate the path ahead.

www.ingramcontent.com/pod-product-compliance
Lightning Source LLC
Chambersburg PA
CBHW070419230526
45471CB00006B/2885